The Rock Cycle

Printed in the United States of America

ISBN 978-0-15-362042-3
ISBN 0-15-362042-0

1 2 3 4 5 6 7 8 9 10   179   16 15 14 13 12 11 10 09 08 07

Harcourt
SCHOOL PUBLISHERS

Visit *The Learning Site!*
www.harcourtschool.com

What Are the Types of Rocks?

VOCABULARY

mineral
rock
igneous
sedimentary
metamorphic

A **mineral** is a solid substance that occurs naturally in rocks or in the ground. Gold is a mineral.

A **rock** is a solid substance that is made up of one or more minerals.

An **igneous** rock forms when melted rock cools and hardens.

A **sedimentary** rock forms when layers of tiny pieces of rock are pressed together.

A **metamorphic** rock is rock changed by heat and pressure into another kind of rock. Marble is a metamorphic rock.

READING FOCUS SKILL

MAIN IDEA AND DETAILS

The main idea is what the text is mostly about. Details tell more about the main idea.

Look for details about different types of rock.

Minerals and Rocks

A **mineral** is a solid substance that is found naturally in rocks or in the ground. Every mineral has special things about it that make it different. These are called its physical properties.

A **rock** is a solid substance made of minerals. A rock can be made of one mineral or many minerals.

Minerals ▼

Fluorite

Calcite

Quartz

▼ This is a mineral from which we get lead.

▲ biotite

▲ mica

▼ Hornblende

▼ Copper comes from this mineral.

There are more than 4,000 minerals. Many minerals look alike. Scientists use physical properties to tell them apart. A mineral's hardness is a physical property. So is whether it is magnetic. Another property is how the mineral reflects light. Another is the color streak it makes when it is rubbed against a tile. All these properties help scientists identify a mineral.

 Tell what properties scientists use to identify minerals.

Granite

Igneous Rocks

Scientists put rocks into three groups. Each group of rocks is formed in a different way.

The first group is igneous rocks. An **igneous** rock forms when melted rock cools and hardens. Melted rock, called magma, cools in different ways. Where and how it cools make igneous rocks different from one another.

Focus Skill **Tell how igneous rocks form.**

Diorite　　　Rhyolite　　　Basalt

Sandstone

Sedimentary Rocks

The second group is sedimentary rocks.
A **sedimentary** rock forms from layers of tiny pieces of
rock. These rock pieces are called sediment.

Wind, water, and ice carry sediment. When they slow
down, sediment drops out. It piles up in layers. The layers
get pressed together. Then water carries minerals through
the layers of sediment. These minerals make the layers
stick together. Over time, they form sedimentary rock.

 Tell how sedimentary rocks form.

Shale

Conglomerate

Metamorphic Rocks

The third group of rocks is metamorphic rocks. A **metamorphic** rock is rock that has changed from another type of rock. Metamorphic rocks can form from any type of rock. This includes other metamorphic rocks.

Slate ▼

Gneiss ▼

Marble ▼

Heat and pressure form metamorphic rocks. Rocks near the surface get pushed down. Pressure squeezes the rock. If the pressure is great enough, the minerals change. Great heat can change minerals in rocks, too. This happens when rocks are pushed deeper into Earth's crust.

 What two processes form metamorphic rocks?

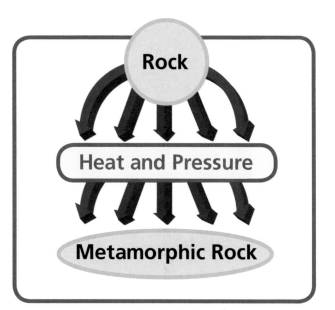

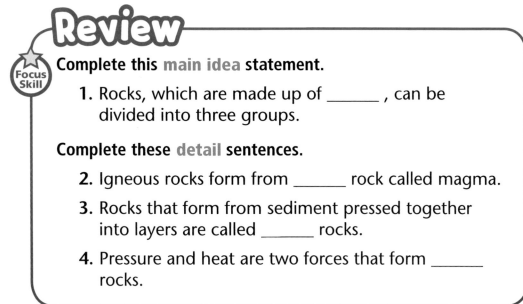

Review

Focus Skill **Complete this main idea statement.**

1. Rocks, which are made up of _____ , can be divided into three groups.

Complete these detail sentences.

2. Igneous rocks form from _____ rock called magma.

3. Rocks that form from sediment pressed together into layers are called _____ rocks.

4. Pressure and heat are two forces that form _____ rocks.

What Is the Rock Cycle?

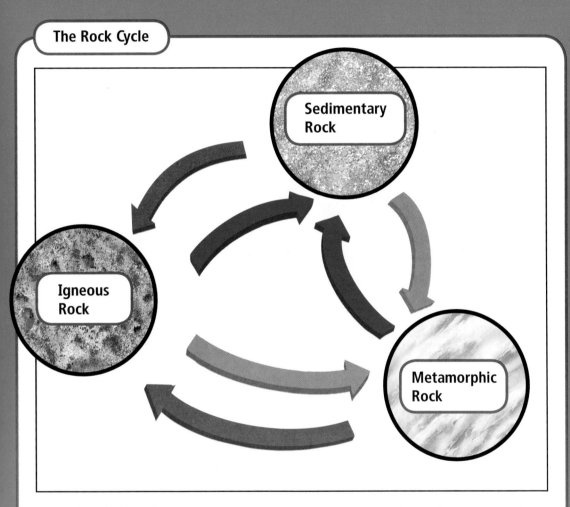

The Rock Cycle

Sedimentary Rock

Igneous Rock

Metamorphic Rock

The **rock cycle** is the sequence of processes that change rocks over long periods. There are many possible paths through the rock cycle.

READING FOCUS SKILL

SEQUENCE

A sequence is the order in which things happen.
Look for the sequence of changes in the rock cycle.

Rock Cycle

Earth's surface is always changing. Forces inside Earth push up mountains. Rain and water wear down mountain rocks. Rivers and wind carry rocks away. Volcanoes erupt, and lava flows. This wears away rock, too.

These changes often happen over thousands or millions of years. They are all part of the rock cycle. The **rock cycle** is the sequence of processes that changes rocks from one type to another.

Rock Cycle

Igneous Rocks

■ = Melting and cooling
■ = Broken down and carried by water, wind, and ice
■ = Pressure and heat

During the rock cycle, each type of rock can be changed into any of the others. The rock cycle can follow many paths. It also happens over and over again.

(Focus Skill) **Is there a first and last step in a rock cycle?**

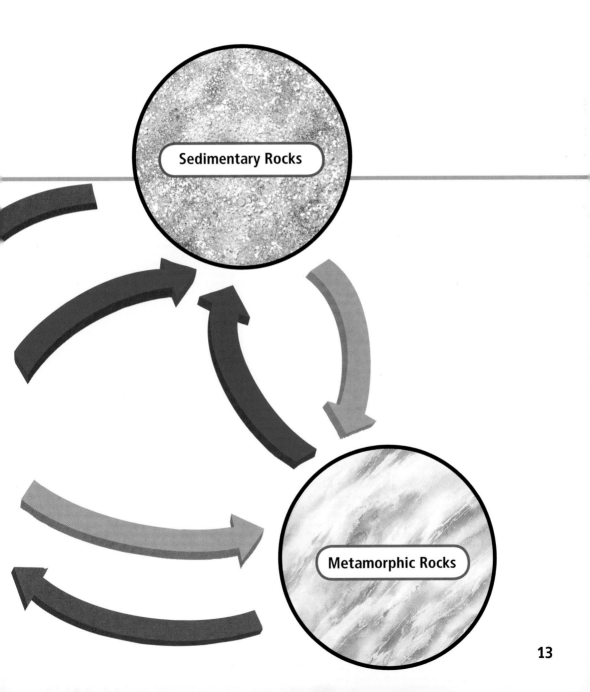

Sedimentary Rocks

Metamorphic Rocks

A Path Through the Rock Cycle

There are many paths through the rock cycle. Here is a description of just one of these paths.

Suppose an igneous rock is pushed up to Earth's surface. Over time this rock breaks down. Layers of this sediment form in water. In time, these layers turn into a sedimentary rock.

▲ Igneous rock may be pushed up to Earth's surface.

▲ Sedimentary rock may form from layers of broken down igneous rock.

▼ Heat and pressure inside Earth may change sedimentary rock into metamorphic rock.

Then suppose this sedimentary rock gets pushed deep into Earth's crust. Heat and pressure change it. It becomes a metamorphic rock.

Then suppose even more heat and pressure act on the metamorphic rock. The rock might melt and then harden. This would form a new igneous rock. Then a new rock cycle begins.

(Focus Skill) Tell what could form after igneous rock breaks down, and the sediment is pressed together in layers.

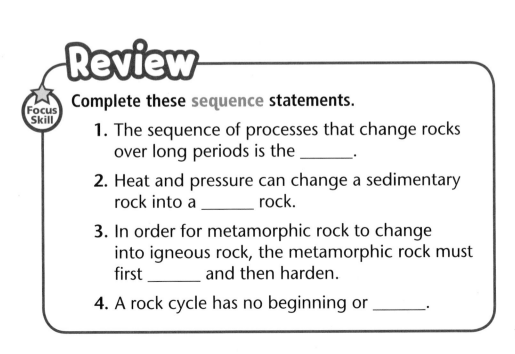

Review

(Focus Skill) Complete these sequence statements.

1. The sequence of processes that change rocks over long periods is the _____.

2. Heat and pressure can change a sedimentary rock into a _____ rock.

3. In order for metamorphic rock to change into igneous rock, the metamorphic rock must first _____ and then harden.

4. A rock cycle has no beginning or _____.

How Do Weathering and Erosion Affect Rocks?

Weathering is the breaking down of rock on Earth's surface into smaller pieces.

Erosion is the process of moving small bits of rock called sediment from one place to another.

Weathering

Weathering is the breaking down of rock on Earth's surface into smaller pieces. Weathering helps shape landforms. It also helps make soil.

There are two types of weathering. One type is chemical weathering. The other type is physical weathering.

Chemical weathering changes the makeup of rock. Some rocks break down when their minerals mix with oxygen. This makes rocks softer and easier to break down.

Rust helps break down this rock. ▶

A sinkhole is the result of physical weathering. It forms when water dissolves underground rocks. This diagram shows how a sinkhole formed.

Focus Skill **Tell the effect of weathering.**

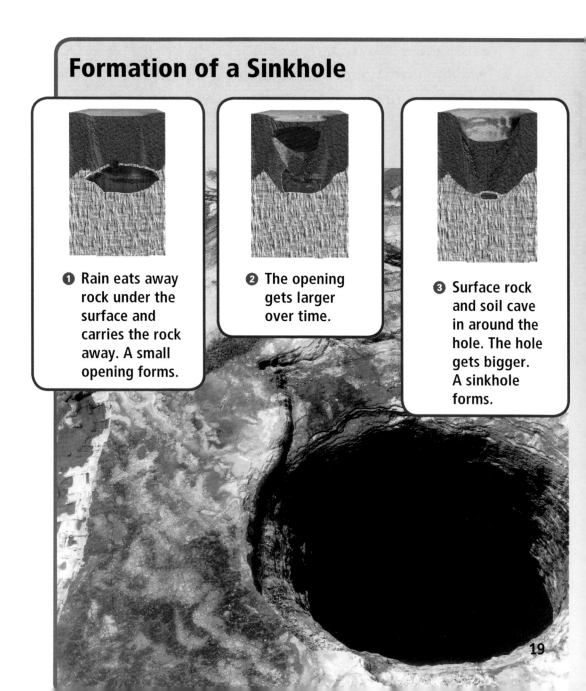

Formation of a Sinkhole

❶ Rain eats away rock under the surface and carries the rock away. A small opening forms.

❷ The opening gets larger over time.

❸ Surface rock and soil cave in around the hole. The hole gets bigger. A sinkhole forms.

Weathering by Other Processes

Physical weathering breaks down rock by changing its shape. Water, ice, living things, and wind cause physical weathering.

Physical weathering causes cracks in sidewalks and other rocks. Water gets in tiny holes in the rock. The water freezes and expands, or gets bigger. This breaks the rock.

Moving water causes weathering. It pushes rocks against each other. Waves also break rocks when they smash into them.

Waves cause weathering.

Rushing water causes weathering. It makes pebbles round and smooth.

Temperature changes can cause weathering. Rock gets bigger when it warms up. It gets smaller when it gets cold. Rocks can break when this happens again and again.

Living things can cause weathering. Tree roots get into cracks. As they grow, they get larger. This can break rocks around them. Animals move rocks near Earth's surface when they dig. These rocks break down more easily.

Wind causes weathering, too. Wind throws sand and soil against other rocks. This chips away at the rocks and breaks them down.

 Tell how living things cause weathering.

Erosion

After weathering happens, erosion often takes place. **Erosion** is the process of moving sediment from one place to another.

Water often causes erosion. Rivers carry sediment. They drop it along their sides or at their mouths. Waves pick up and drop sand.

Wind is another cause of erosion. Wind picks up sediment. It can drop sand into big piles called sand dunes.

Water caused erosion of the river's banks. ▼

Wind can blow away dry, loose soil.

Glaciers also cause erosion. They scrape the ground as they move. Glaciers pick up rocks and soil. They erode mountains and other landforms. Melting glaciers leave behind water and huge piles of rocks. Glaciers have shaped much of the northern United States.

 What are the three main causes of erosion?

◀ Glaciers cause erosion.

Focus Skill **Complete these cause and effect statements.**

1. Wind, water, and plants can all cause _____.

2. Tree _____ can get into cracks and then break rocks.

3. Wind causes _____ when it picks up sand and then puts it down in a new place.

4. _____ can change the shape of landforms by picking up rocks and soil as they slowly move over the land.

What Is Soil?

VOCABULARY

horizon
bedrock
humus
sand
clay

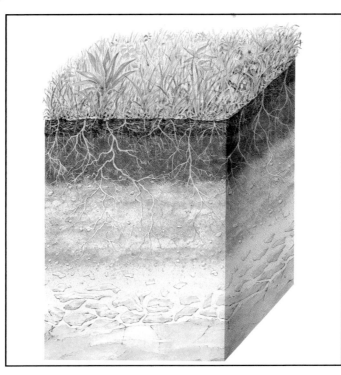

Most soil has several **horizons**, or layers. The bottom horizon is bedrock. **Bedrock** is the solid rock that forms Earth's surface.

Humus is the remains of dead plants and animals found in soil.

Sand particles are the largest particles that make up soil.

Clay particles are the smallest particles that make up soil.

READING FOCUS SKILL
COMPARE AND CONTRAST

When you compare and contrast, you tell how things are alike and different.

Look for ways to compare and contrast soils.

What Is in Soil

Soil is made up of different things. It has living and nonliving parts. Almost half of soil is sediment, or weathered rock. Most soil contains humus, too. **Humus** is the remains of dead plants and animals. Living things such as worms and insects live in soil. Large amounts of water and air are also in soil.

How Soil Forms

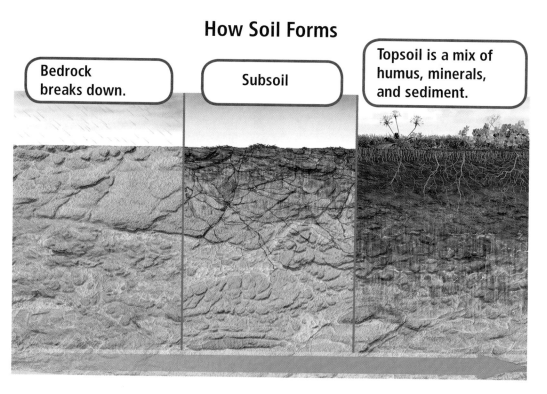

Bedrock breaks down.

Subsoil

Topsoil is a mix of humus, minerals, and sediment.

Most soil has **horizons**, or layers. Some soils have many horizons. Others have just a few.

Soil horizons are alike in some ways. The top layer is called topsoil. It often has humus. Lower layers have partly weathered rock. More minerals are found in these layers. The bottom layer is called bedrock. **Bedrock** is the solid rock that forms Earth's surface.

Soils take a long time to form. That is one reason soil is an important resource.

 Tell how the top and bottom horizons of soil are different.

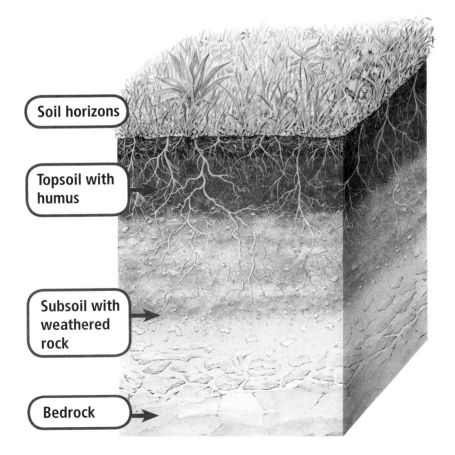

Soil horizons

Topsoil with humus

Subsoil with weathered rock

Bedrock

Types of Soil

Scientists group soils by looking at their physical properties. One property is particle size. Each type of soil is a mix of particles of different sizes. The largest particles are **sand**. The smallest are **clay**. Silt particles are in between sand and clay in size.

Another property is soil texture. Texture is the way soil feels. Sandy soils feel rough. Clay soils feel smooth.

The amount of water soil holds is a third property. Clay soils hold a lot of water. Sandy soils do not.

Types of Soil

Sandy soil · Fertile soil · Clay soil

Soil Horizons in Three Different Places

Grassland soil

Desert soil

Florida soil

Soil type depends on where soils form. Dark soils may form where there is a lot of humus, such as in a forest. Light soils may form where there is little humus, such as in a desert.

 Tell how desert and forest soils might be alike and different.

Soil and Plants

All plants need soil. They get nutrients, water, and oxygen from soil.

Some soils are fertile. This means they have a lot of nutrients. Plants grow well in these soils. Other soils have fewer nutrients. Plants do not grow as well in these soils.

People add nutrients to make soils better. The nutrients are in fertilizers. Some fertilizers are made of chemicals. Others are natural. Animal waste and decayed plants are natural fertilizers.

◀ Natural fertilizer

Some soils hold water better than others. When soil is too dry, plants cannot live. People add water to soil so plants can grow there.

 Compare and contrast the two kinds of fertilizer.

Chemical fertilizer ▶

Review

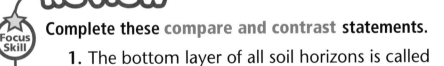

Complete these compare and contrast **statements.**

1. The bottom layer of all soil horizons is called
   _____.

2. Sand has the _____ particles and clay has
   the _____.

3. Fertile soils have more _____ than less
   fertile soils.

4. Clay soils feel smooth and sandy soils feel _____.

GLOSSARY

bedrock (BED•rahk) The solid rock that forms Earth's surface

clay (KLAY) The smallest particles that make up soil

erosion (uh•ROH•zhuhn) The process of moving sediment from one place to another

horizon (huh•RY•zuhn) A layer in the soil

humus (HYOO•muhs) The remains of decayed plants or animals in the soil

igneous rock (IG•nee•uhs RAHK) A type of rock that forms from melted rock that cools and hardens

metamorphic rock (met•uh•MAWR•fik RAHK) A type of rock that forms when heat or pressure changes an existing rock

mineral (MIN•er•uhl) A solid substance that occurs naturally in rocks or in the ground

rock (RAHK) A solid substance made up of one or more minerals

rock cycle (RAHK CY•kuhl) The sequence of processes that change rocks from one type to another over long periods

sand (SAND) The largest particles that make up soil

sedimentary rock (sed•uh•MEN•ter•ee RAHK) A type of rock that forms when layers of sediment are pressed together

weathering (WETH•er•ing) The breaking down of rocks on Earth's surface into smaller pieces